Monica Torres Barrios
Carlos A. Mendoza Jacomino

UNIVERSITY SUSTAINABILITY

Monica Torres Barrios
Carlos A. Mendoza Jacomino

UNIVERSITY SUSTAINABILITY

ALEJANDRA'S JOURNEY

ScienciaScripts

Imprint

Cover image: www.ingimage.com

This book is a translation from the original published under ISBN 978-613-9-41207-5.

Publisher:
Sciencia Scripts
is a trademark of
Dodo Books Indian Ocean Ltd. and OmniScriptum S.R.L publishing group

120 High Road, East Finchley, London, N2 9ED, United Kingdom
Str. Armeneasca 28/1, office 1, Chisinau MD-2012, Republic of Moldova, Europe
Managing Directors: Ieva Konstantinova, Victoria Ursu
info@omniscriptum.com

Printed at: see last page
ISBN: 978-620-8-52715-0

Index

Introduction

In recent years, sustainability has established itself as a key principle in the fight against global environmental challenges. The need to protect our natural resources and reduce the impact of human activity on the planet has never been more urgent. In this context, this book narrates the transformative journey of Alejandra, a young industrial engineering student who, upon entering university, begins to connect the knowledge she has acquired with a deep commitment to sustainability.

Through the pages of *University Sustainability: Alejandra's Journey*, the reader will enter into the experiences of a student who, from her first steps at university, questions her role in the environmental problems that affect not only her country, Nicaragua, but the whole world. Influenced by her studies in sustainability and the environment, Alejandra decides to embark on a process of both personal and collective change, starting with small, everyday actions aimed at reducing her ecological footprint. However, she soon realises that true transformation requires a collective approach, collaboration with other students, teachers and local communities.

This book documents not only Alejandra's efforts to reduce plastic consumption, improve energy efficiency in her home and promote sustainability at her university, but also her involvement in wider initiatives. Through her participation in composting projects, the creation of a community garden and the proposal of a renewable energy innovation centre, Alejandra manages to inspire those around her and, in doing so, initiate change in her immediate environment. Throughout this process, she discovers the power of collaboration, the importance of environmental education and the need to integrate sustainability into all areas of daily life.

With a comprehensive approach that encompasses the circular economy, renewable energies and the responsible management of natural resources, this book offers an in-depth reflection on how universities can be key spaces for fostering commitment to the environment. It also highlights how local actions, however small, can have a significant impact if carried out with vision and dedication.

In this story, readers will find not only a story of personal transformation, but also a call to action. Alejandra, like other young people committed to change, demonstrates that a more sustainable future is possible, as long as each of us is willing to do our part and work with others to make a positive and lasting impact.

Foreword

Sustainability has become a global imperative that transcends disciplines and sectors, challenging us to rethink our actions and responsibilities towards the environment and society. In this context, University Sustainability: Alejandra's Journey is presented as a work that explores, from an academic and practical perspective, the impact of individual decisions on collective transformation towards a more sustainable future.

This book chronicles the journey of Alejandra, a university student whose personal and academic journey illustrates how the incorporation of sustainability principles can bring about profound change in diverse environments. From her first experiences in the classroom to the implementation of high-impact projects in her community and university, Alejandra demonstrates how knowledge, commitment and strategic action can converge to address the most pressing challenges of our time.

Organised into thematic chapters, the text combines inspirational narratives with thoughtful analysis, ranging from the fundamentals of sustainability to the implementation of concrete initiatives at university and community levels. Each chapter is designed to provide the reader with a comprehensive understanding of the challenges and opportunities involved in working towards sustainability, highlighting the importance of interdisciplinary collaboration, environmental education and transformative leadership.

Alejandra's Journey is not just a personal account; it is an invitation to critical reflection and collective action. This book offers a rigorous approach that articulates key concepts, case studies and practical examples, making it a valuable tool for both students and academics and practitioners interested in integrating sustainability into their projects and activities.

In its pages, it underlines the relevance of educational institutions as agents of change, as well as the role of individuals in building sustainable development. This book aims to inspire not only immediate action, but also the adoption of a long-term perspective, in which every decision becomes a significant contribution to global well-being.

We invite the reader to immerse themselves in this transformative story and to consider how the ideas and learnings captured in this book can be applied in their own context. University Sustainability: Alejandra's Journey is a reminder that change, though complex, is possible when will, knowledge and action come together.

CAPITULO 1

Chapter 1: Introduction to Sustainability

From the beginning, Alejandra has always had a special connection to the environment. She grew up surrounded by Nicaragua's nature, and her parents often took her to explore nearby nature reserves and parks. However, it was during her final years of high school that she began to hear about major global challenges, such as climate change, ocean pollution and biodiversity loss. It was then that she began to realise that these problems were not abstract or distant, but also affected her country and her community.

Prepared by own source

1.1 Sustainability in university life

Upon arriving at the American University of Managua, Alejandra was excited to begin her studies in Industrial Engineering. Her main goal was to learn how to optimise processes and create efficient systems, but it wasn't until she enrolled in the Sustainability and Environment class that she really understood the connection between her career and the need to protect the planet. The class, taught by Professor Morales, covered topics such as the circular economy, renewable energy and waste management, all of which resonated deeply with Alejandra.

Professor Morales was an inspirational figure. She spoke passionately about the importance of sustainability and the need for a profound change in the way things were done. She explained that sustainability was not only an environmental concern, but a principle that should be integrated into all areas of life and the economy. Alejandra was motivated by these ideas and began to question how she could apply these concepts in her daily life and in her future career as an engineer.

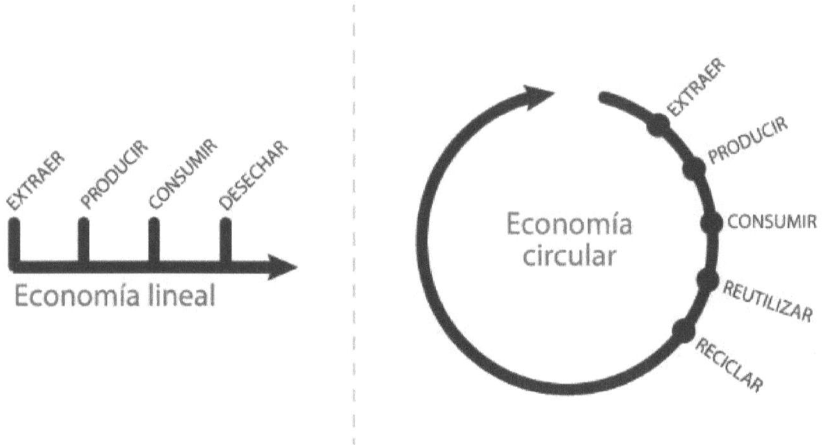

Prepared by own source

One of the first concepts that caught Alejandra's attention was the circular economy. Professor Morales taught them that, unlike the linear economic model, where resources are extracted, used and then discarded, the circular economy seeks to keep materials in use for as long as possible. This involved redesigning products to be more durable, recoverable and recyclable. Alejandra realised that this approach not only made sense from an environmental point of view, but could also be an opportunity to innovate and create new business models.

1.2 The magnitude of the environmental challenges

As the lessons progressed, Alejandra began to better understand the magnitude of the environmental problems facing the world. Professor Morales presented alarming data on the loss of biodiversity, the depletion of natural resources and the effects of climate change in vulnerable regions such as Central America. Alejandra couldn't help but feel overwhelmed, but she also felt a growing urge to do something about it. She knew that her career gave her the tools to optimise processes and improve efficiency, but now she understood that she had to do it in a sustainable way.

One of the lessons that struck Alejandra the most was about the importance of renewable energy. Nicaragua, with its abundant sun and wind, had great potential for clean energy development, and Professor Morales showed them examples of projects already underway in the country. Alejandra was inspired by the idea that her own country could lead the transition to a more sustainable energy model and began to investigate further how industrial engineering could contribute to this goal.

In the midst of all this learning, Alejandra began to have many questions: How feasible was it really to make a change from the university? How could she, as a student, contribute meaningfully to sustainability? These questions led her to reflect on her own role and the individual actions she could take. She decided that she didn't just want to stick to theoretical knowledge; she wanted to put it into practice and make a difference in her immediate environment.

The university organised a series of conferences on the UN Sustainable Development Goals (SDGs), and Alejandra attended all of them. The SDGs were a series of global goals to eradicate poverty, protect the planet and ensure peace and prosperity for all. Alejandra was particularly drawn to the goals related to affordable and clean energy , climate action and responsible production and consumption. She understood that sustainability was a collective effort and that each person, from his or her place, could contribute to these global goals.

The Sustainability class not only gave Alejandra technical knowledge about environmental problems, but also motivated her to act. During one of the sessions, Professor Morales assigned them a group project in which they had to analyse the environmental impact of the university and propose solutions to reduce it. Alejandra joined a group of peers who shared her concerns, and together they began researching the use of plastics, energy consumption and waste management on campus.

Working on this project was a turning point for Alejandra. She realised that, although the environmental problems seemed overwhelming, there were many things that could be done locally to make a difference. Together with her team, she submitted a proposal to reduce the use of single-use plastics in the university cafeteria and to implement composting stations where food scraps could be composted for the campus garden. This experience showed him that change was possible if you work as a team and if there is a will to act.

One of the most significant moments during this process was when Alejandra's group presented their proposal to the university administration. Although there was some initial resistance, Professor Morales supported them and helped them refine their arguments to demonstrate the benefits of the proposed measures. Eventually, the administration agreed to implement some of the suggestions, and Alejandra was deeply gratified to see her ideas begin to become a reality.

From that experience, Alejandra decided that she wanted to get more involved in sustainability projects on campus. She joined a student group working on environmental initiatives and started attending meetings and activities. Gradually, she began to meet other students who were passionate about the environment and together they started to come up with new projects. It was at one of these meetings that the idea of the community garden was born, a space where students could learn about sustainable agriculture and connect with nature.

The community garden was one of the first big projects Alejandra became actively involved in. Together with her peers, she began to look for funding and support within and outside the university. Although they faced many challenges at the beginning, the group's persistence and commitment to the cause allowed them to obtain the necessary resources to get the project off the ground. Alejandra learned a lot during this process, not only about agriculture, but also about the importance of collaboration and resilience.

While working in the community garden, Alejandra began to explore other ways to promote sustainability at the university. The idea of a renewable energy innovation centre began to form in her mind, and although she knew it would be an ambitious project, she couldn't help but imagine the impact it could have. He started talking about the idea with some of his professors and colleagues, and was surprised to find that many shared his enthusiasm. It was then that he decided that this would be one of his next goals.

Alejandra also began to reflect on the importance of education in sustainability. She realised that many students were unaware of the impact their daily actions had on the environment, and decided she wanted to change that. Together with the sustainability group, she organised workshops and activities to teach her peers about recycling, composting and reducing energy consumption. These workshops not only informed the students, but also helped to create a community that was more aware and committed to change.

One of the aspects Alejandra enjoyed most during this process was the teamwork. She discovered that when people come together for a common goal, they can achieve incredible things. The collaboration with her peers and teachers motivated her to keep going, even when faced with obstacles. Each small achievement, such as installing a composting station or reducing the use of plastics in the cafeteria, was a sign that they were on the right track and that their efforts were having a real impact.

1.3 Beyond the campus: Community initiatives

Alejandra also started to get involved in off-campus activities. She joined a volunteer group working on reforestation projects on the outskirts of Managua and participated in river and beach clean-up campaigns. These experiences allowed her to see first-hand the environmental challenges facing her country, but also filled her with hope as she saw how people were willing to work together to care for the environment. Alejandra realised that sustainability was not just an academic concept, but a practice that should be integrated into all aspects of life.

As time went on, Alejandra realised that her vision of sustainability had evolved. At first, she thought it was simply about reducing negative impact on the environment, but now she understood that sustainability was also about creating stronger, fairer and more resilient communities. This realisation led her to want to keep learning and to seek new opportunities to expand her knowledge and impact.

It was then that the opportunity arose to apply for a scholarship to study abroad. One of her professors, impressed by her dedication and leadership in sustainability projects, recommended her for a scholarship at a university that had an advanced programme in renewable energy. Alejandra did not hesitate and applied, and to her surprise and joy, she was accepted. This opportunity would allow her to learn from experts in the field and bring that knowledge back to Nicaragua to continue working for a more sustainable future.

The day she received the news of the scholarship, Alejandra was excited and a bit nervous. She knew it would be a great challenge, but also a unique opportunity to grow and learn. Before leaving, she made a commitment to her colleagues to continue supporting them in the projects they had worked on together and to return with new ideas and tools to continue promoting sustainability at the university and in the community. She knew the journey would not be easy, but she was ready to face whatever came and to continue working for change.

CAPITULO 2

Chapter 2: Alejandra: A New Phase

2.1 Setting out on a new path

Alejandra has just finished high school. Nineteen years old and with the characteristic enthusiasm of someone who is starting a new stage, Alejandra arrives at the American University campus, full of expectations and doubts. She is attracted by the idea of solving complex problems, improving systems and, above all, contributing to the well-being of her community. In her first semester, she has the opportunity to enrol in a class called "Sustainability and the Environment". At first, she doesn't have high expectations for how this class might change her perspective, but something in her begins to resonate.

The Sustainability class is taught by Professor Morales, a woman who is passionate about the environment and community initiatives. During the first few sessions, the teacher introduces concepts about the circular economy, renewable energy and the importance of waste reduction. Alejandra listens attentively, but can't help but feel that it all sounds very theoretical. "How can I really do anything?" she often asks herself. The frustration of not knowing how to apply these concepts to her daily life leads her to question whether she can really make a difference.

Despite her doubts, Alejandra decides to get more involved in the class. She participates in discussions, asks questions and begins to notice that some of her classmates share her concerns. Little by little, a small group of students interested in exploring sustainability issues in more depth begins to form. Among them, Alejandra finds a source of motivation and support.

University life also brings new challenges for Alejandra. Adapting to a different environment, meeting new people and finding a balance between her studies and her personal life is not easy. However, the idea of being able to contribute to something bigger drives her on. In one class, Professor Morales invites students to participate in a research project on the impact of food waste on campus. Alejandra is intrigued and decides to join the project.

Working on the research project allows Alejandra to see how the concepts she learns in class apply in real life. Together with her classmates, she conducts surveys and analyses data on food waste in the university cafeteria. They discover that a large amount of food is wasted every day, and that many people are unaware of the environmental impact of their habits. Alejandra is motivated to find solutions and starts working on proposals to reduce waste.

As she becomes more involved in the project, Alejandra realises that sustainability is not only a matter of changing individual habits, but also of influencing the policies and practices of institutions. Together with her colleagues, she submits a proposal to the university administration to implement a composting programme in the cafeteria. Although they face resistance at first, their persistence and determination finally get the proposal approved.

The success of the composting project gives Alejandra a new perspective on what is possible when working as a team. She begins to see her peers not just as classmates, but as allies in the fight for a more sustainable future. This experience also teaches him the importance of communication and collaboration. Without the support of his classmates and the university administration, the project would not have been possible.

Alejandra also begins to explore other areas of sustainability that interest her. She joins a group of students working on a solar energy project for the campus. Although she has no experience in the subject, she is willing to learn and contribute what she can.

Prepared by own source

Working on this project allows him to meet students from other careers, such as electrical engineering and architecture, and to learn about the importance of renewable energy in the transition to a more sustainable future.

Participating in these projects also gives Alejandra the opportunity to develop skills she had not considered before. She learns to work in a team, to manage projects and to communicate effectively with different people. These skills are not only useful for her studies, but also give her confidence in herself and her ability to make a difference.

Over time, Alejandra begins to notice changes in her way of thinking and acting. She no longer sees sustainability as something abstract or distant, but as something she can apply in her daily life. She starts to make small changes in her routine, such as reducing plastic consumption, saving energy and water, and opting for more sustainable products. Although at first these changes seem insignificant to her, she soon realises that they have a real impact, both on her life and on the lives of the people around her.

The sustainability class also allows Alejandra to meet teachers and experts working in the field. On one occasion, Professor Morales invites a climate change expert to give a talk on the effects of global warming in Central America. The talk has a profound impact on Alejandra, as it shows her the real and tangible consequences of climate change in her own region. This motivates her even more to continue working for a more sustainable future.

2.2 Beyond the classroom

Alejandra also starts to get involved in activities outside the classroom. She joins a volunteer group that works on cleaning up beaches and rivers in Managua. These activities allow her to see first-hand the impact of pollution and the importance of keeping ecosystems clean. It also gives her the opportunity to meet others who share your passion for the environment and to feel part of a community committed to change.

As time goes by, Alejandra realises that her vision of sustainability has evolved. At first, she thought it was only about caring for the environment, but now she understands that it is a much broader concept that includes social justice, economic equity and the well-being of all people. This new understanding leads her to want to learn more and to look for ways to apply this knowledge in her Industrial Engineering career.

CAPITULO 3

Chapter 3: Unanswered Questions

3.1 Questioning the impact of individual actions

During one of the Sustainability classes, Alejandra listens to the teacher talk about the impact of small individual actions on the environment. "We can make a difference," the teacher says with conviction. But Alejandra can't help wondering if that's true: what's the point of reducing plastic use if there's still tons of rubbish around her? That afternoon, she decides to go to the campus café, where she meets up with a couple of friends. The warm atmosphere and the aroma of freshly brewed coffee relax her, but her mind is still restless.

"I think the teacher is right, but I don't know how to start," says Alejandra as she stirs her cup of coffee. Her friend Carlos, who is also in the class, nods. "Yes, I understand what you are saying. It's hard to think that our actions can have a real impact when everything seems so big," he replies. The conversation turns deep, and the two share their frustrations and hopes for the future. Although they don't find clear answers, Alejandra feels she is not alone in her concerns.

That evening, when Alejandra got home, she thought about the words of her friend Lucía, who was also in the café. "Maybe we can't change the world on our own, but we can start small," she said with a smile. That phrase stuck in Alejandra's mind and drives her to look for ways to start, even if they are small.

3.2 Transforming concerns into initiatives

Alejandra decides to make a list of small actions she could implement in her daily life to be more sustainable. From reducing the use of plastic to saving energy and water at home, the list gets longer and longer as she researches. He realises that although each individual action may seem insignificant, together they can add up to a big impact. The idea of being able to inspire others to do the same gives her new energy.

Prepared by own source

In class, Professor Morales assigns them a group project to investigate how sustainability could be improved on campus. Alejandra, along with Carlos and Lucía, decides to focus on reducing the use of plastics. They work on an awareness campaign to promote the use of reusable bottles and reduce the consumption of single-use plastics in the cafeteria. Although they are initially unsure about how the campaign will be received, they soon realise that many students are willing to support the initiative.

The campaign is starting to gain traction, and Alejandra feels a great sense of satisfaction that her efforts are having an impact. They organise workshops and talks on the importance of reducing the use of plastics and offer practical alternatives for students. Little by little, more people join the cause, and Alejandra realises that change is possible with teamwork and patience.

As they move forward with the project, Alejandra and her peers face challenges. Some people are not interested in changing their habits, and there are times when they feel that their efforts are not paying off. However, each small achievement, such as seeing a student using a reusable bottle or hearing someone talk about the campaign, gives them the motivation they need to keep going.

Alejandra also begins to question other aspects of her life. She realises that many of the things she consumes are not necessary and that she could reduce her environmental impact by changing her consumption habits. She decides to start shopping more consciously, opting for local and sustainable products whenever possible. Although at first this involves extra effort, he soon realises that it is a way of supporting his community and contributing to the care of the planet.

Professor Morales also assigns them readings on the impact of climate change in different parts of the world. Alejandra is particularly moved by the stories of communities who have lost their homes due to natural disasters. These readings make her reflect on the importance of acting now to prevent more people from suffering the consequences of climate change. She realises that although her actions may seem small, they are part of a global effort to protect the planet.

3.3 Leading change through collaboration

One day, during one of the sustainability group meetings, Alejandra comes up with the idea of organising an event to make more students aware of the importance of sustainability. The idea is to do something bigger than the workshops and talks they have been doing. She wants to organise a day of activities on campus, with information stands, practical workshops and talks by experts. His classmates are enthusiastic about the idea and start planning the event.

The planning process is not easy. It requires coordinating with different university departments, getting permissions, contacting experts and organising the activities. Alejandra feels overwhelmed at times, but she also realises that she is learning a lot. She is developing organisational, leadership and communication skills she never imagined she would need. She also feels great satisfaction in seeing how her idea is taking shape.

Prepared by own source

The day of the event finally arrives, and Alejandra can't help but feel nervous. She wonders if people will attend, if the activities will be well received and if they will get the message across. To her surprise, the event turns out to be a success. Many students attend, participate in the workshops and are interested in learning more about sustainability. Alejandra is proud of what they have achieved and realises that, although there is still a long way to go, they have taken a big step in the right direction.

After the event, Alejandra and her peers meet to reflect on what they have learned and how they can move forward. They decide that they want to continue organising activities and that they want to involve more people. Alejandra feels that she has found a purpose at university and that, although the road is not always easy, it is worth continuing.

CAPITULO 4

Chapter 4: Exploring Options

4.1 A step towards collective action

Determined to go beyond the words and theories in class, Alejandra starts researching on her own. She consults articles, documentaries and talks to some teachers who are involved in environmental projects. She discovers that there are student groups dedicated to sustainability, and although she is a little intimidated at first, she decides to attend one of their meetings. The students are organising activities to reduce waste on campus, and Alejandra begins to see a glimmer of hope.

At the first meeting he attends, he meets Marcos, a biology student who is leading the group. Marcos is enthusiastic and has many ideas on how to improve waste management on campus. Alejandra is inspired by his passion and decides to get involved in one of the projects: a campaign to reduce the use of single-use plastics at the university. Together, they organise workshops to raise awareness among students about the impact of plastic and propose reusable alternatives.

Prepared by own source

Alejandra also begins to notice how her peers react to her efforts. Some are supportive and join in the activities, while others seem indifferent. Although she sometimes feels discouraged, small achievements, such as seeing her friends using reusable bottles, give her the strength to keep going.

As she becomes more involved in the student group, Alejandra realises that there are many ways she can contribute. She participates in campus clean-up activities, organises sustainability talks and collaborates with other students to create a community garden. Each of these activities allows her to meet more people who share her passion for the environment, and Alejandra begins to feel like she is part of a community.

4.2 Beyond the campus: Community projects and reforestation

Alejandra also begins to explore other initiatives outside the university. She joins a group of volunteers working on a reforestation project on the outskirts of Managua. Every weekend, she meets with the group to plant trees and restore degraded areas. Although the work is hard and physically demanding, Alejandra feels great satisfaction in seeing the landscape begin to change. Reforestation not only helps combat climate change, but also provides habitats for local wildlife and improves air quality.

During one of the sustainability group meetings, Marcos proposes the idea of collaborating with a local organisation working on waste management. The organisation focuses on educating communities on how to reduce, reuse and recycle waste. Alejandra is enthusiastic about the idea and decides to join the project. Together, they organise workshops in different communities in Managua, teaching people how to separate waste and how to compost at home. Alejandra is surprised to see people's interest and willingness to learn.

4.3 Education as an engine for change

As she gets involved in these projects, Alejandra realises the importance of education in sustainability. Many people are unaware of how their daily actions affect the environment , and lack of information is one of the biggest obstacles to change. Alejandra decides that she wants to focus on environmental education and starts working on creating teaching materials that can be used in schools and communities. Together with Marcos and other colleagues, they develop brochures and videos that explain in a simple way concepts such as recycling, waste reduction and the importance of conserving natural resources.

One of the most significant moments for Alejandra occurs during a visit to a school in a vulnerable neighbourhood in Managua. Together with her group, they organise an activity to teach the children about the importance of taking care of water. They use games and dynamics to make the activity fun and educational. Alejandra is moved by the children's enthusiasm and realises that they are the future, and that educating them is fundamental for long-term change.

CAPITULO 5

Chapter 5: Managua and the Quest for Change

5.1 Inspiration in communities: Urban gardens and recycling

Alejandra also wants to see what is happening beyond the university walls. She joins a small group that visits community projects in different neighbourhoods of Managua, where she learns about recycling initiatives and urban gardens. These visits help her to better understand the difficulties communities face in implementing sustainable change, but also show her the power of organisation and collective will.

Alejandra's first visit is to a neighbourhood called Los Laureles, where a community has managed to establish a small urban garden in the middle of the empty land between several houses. Despite the scarcity of resources, the residents have managed to grow vegetables and herbs that they use for their own consumption and also to share among their neighbours. Alejandra is impressed by the creativity and dedication of the people leading this project. Doña Marta, one of the garden's founders, explains that they started with donated seeds and recycled materials to create the garden's fencing, and Alejandra is deeply inspired by their determination.

Alejandra realises that sustainability does not only depend on advanced technology or large investments, but also on the ability of people to work together towards a common goal. During the visit, she also meets other young people who have become involved in the project and who have seen positive changes in their community thanks to the garden. Children from the neighbourhood run around the area, helping to water the plants and playing on the land where before there was only waste. For Alejandra, seeing this kind of transformation is encouraging and reaffirms her desire to get more involved in similar initiatives.

5.2 Learning from community resilience

On another visit, Alejandra meets a group of women working on a plastic recycling project.

These women collect bottles and other plastic waste they find in the streets of their community and take them to a recycling centre where they are processed into reusable materials. The project has allowed them to generate a small extra income, which has been crucial for many of them. Alejandra talks to Doña Rosa, who tells her how recycling has allowed her to buy school supplies for her children. The resilience and creativity of these women impresses Alejandra, who begins to better understand how sustainability also has a positive economic and social impact.

Alejandra also visits a community centre where people are taught how to compost organic waste. There she meets Pedro, an enthusiastic young man who leads workshops to teach families how to transform food waste into compost for their gardens and allotments. Pedro tells her that, at first, many neighbours were sceptical about the idea of composting, but little by little they saw the benefits. Alejandra is encouraged to learn more about the process and thinks about how she could bring the practice to her community and university.

During these visits, Alejandra begins to understand the urban realities of Managua from a different perspective. Sustainability is not only about individual initiatives, but also about strengthening the social fabric of communities. Often, environmental problems in Managua are not due to a lack of interest on the part of the people, but to a lack of opportunities and resources. Alejandra is moved to see how, despite the difficulties, communities find ways to organise themselves and improve their living conditions.

Alejandra also takes part in a tour with a group of students from other universities who have come together to learn about community initiatives in Managua. In one of the neighbourhoods, they meet a group of young people who have organised a campaign to clean up a nearby river. The river, which used to be full of rubbish and debris, is now flowing cleaner thanks to the work of these volunteers. Alejandra joins the effort, and although the work is hard, she feels a great sense of satisfaction at seeing the tangible change they are making.

Prepared by own source

As she becomes more involved in these activities, Alejandra realises that change is not always immediate, but that every small step counts. She is inspired by people's ability to meet challenges with creativity and solidarity. The connection that is created between community members when they work together is something Alejandra has not experienced before. This motivates her to keep going and to look for ways to connect her studies in industrial engineering with the needs of the community.

5.3 Environmental education: a transformative tool

Another important aspect Alejandra learns during her visits is the importance of environmental education in the communities. Many of the people she talks to tell her that before they did not know how their daily actions could affect the environment. Thanks to the workshops and the information shared by the community leaders, they have been able to make changes in their habits that benefit both their families and the environment. Alejandra begins to see education as a fundamental tool for sustainability and decides that she wants to focus her efforts on bringing this knowledge to more people.

One afternoon, while visiting a school in a neighbourhood in Managua, Alejandra and her group organise an activity to teach the children about the importance of taking care of water. They use games and dynamics to make the activity fun and educational. The children participate enthusiastically, and Alejandra is moved by the impact that education can have on the younger generation. She understands that children are the future and that educating them from an early age is essential for lasting change.

Prepared by own source

Alejandra also realises that community initiatives not only have an environmental impact, but also help to strengthen ties between neighbours. In one of the neighbourhoods, she meets Don Manuel, an old man who has lived there all his life. Don Manuel tells her how, thanks to the community garden, he has got to know his neighbours again and felt useful by helping with the cultivation tasks. Alejandra understands that sustainability also has a very important human component: it is about building relationships and working together towards a common goal.

5.4 Innovation and policies for a sustainable future

Alejandra's experiences in Managua also make her reflect on the role of public policy in sustainability. Many of the communities she visits do not have the necessary support from the government to implement their projects, and this limits the scope of their efforts. Alejandra begins to research environmental policies in Nicaragua and realises that although laws and programmes exist, they are often not effectively implemented. This leads her to consider getting involved in public policy advocacy that supports community initiatives and promotes sustainability.

One of the initiatives that impacts Alejandra the most is a solar energy project in a rural community near Managua. Thanks to the installation of solar panels, families in the community have been able to access electricity for the first time. Alejandra is excited to see how a relatively simple technology can have such a big impact on people's lives. This motivates her to continue learning about renewable energy and to think about how she could apply this knowledge in her future projects.

During one of the visits, Alejandra also has the opportunity to meet a group of craftswomen who use recycled materials to create products that they sell in local markets. The women show her how to transform plastic bags and other waste into colourful bags and accessories. Alejandra is impressed by the skill and creativity of these craftswomen, and realises that sustainability also means finding innovative ways to give value to what others consider waste.

Alejandra feels more and more connected to the people she meets on these visits. She realises that sustainability is not just an academic concept, but a reality that is lived every day in communities. The stories of resilience and creativity she hears motivate her to keep going and to look for ways to contribute.

CAPITULO 6

Chapter 6: Personal Actions

6.1 Reducing environmental impacts in everyday life

Inspired by all that she has learned and seen, Alejandra decides that it is time to start acting from what is closest to her: her own life. Although the idea of changing the world seemed overwhelming at first, the prospect of starting with small, everyday actions seems more achievable. So she decides to implement some changes in her home and daily routine that can help reduce her environmental impact. Her first step is to reduce the use of plastics. Alejandra starts by eliminating plastic bags, opting for reusable cloth bags when she goes to the market. She also replaces plastic bottles with a stainless steel bottle that she carries with her everywhere she goes. At first, these changes seem small, but Alejandra realises that every conscious choice is a step towards a more sustainable future.

Prepared by own source

In addition, Alejandra decides to make adjustments to her diet. She starts to reduce meat consumption and to include more plant-based foods. She discovers that the livestock industry has a big impact on the environment due to greenhouse gas emissions and the intensive use of natural resources . Although she finds it difficult to adapt at first, she gradually begins to enjoy new recipes based on vegetables and legumes. Her younger sister, interested in what Alejandra is doing, joins her in the kitchen, and together they begin to prepare healthy and sustainable dishes. Alejandra is pleased to see how her family is also interested in making small changes.

Alejandra is also interested in reducing food waste in her household. She discovers that a lot of food goes to waste each year and decides that she can do her part by making better use of the food she buys. He starts planning meals for the week, reusing leftovers to prepare new dishes and storing food properly to prevent it from spoiling. She also creates a small compost pile in the garden to transform food scraps into compost for the plants. She finds this process fascinating and, by seeing how organic waste becomes a valuable resource, Alejandra feels more connected to nature.

Prepared by own source

One of the most significant changes Alejandra makes in her life is to reduce her energy consumption. She starts to be more conscious of electricity use in her home, turning off lights and unplugging electronic devices when she is not using them. She also replaces conventional light bulbs with energy-saving LED bulbs, and suggests to her family to use the washing machine only when there are enough clothes to fill a full load. Although these are small actions, Alejandra realises that, over time, these habits can make a big difference to household energy consumption.

Alejandra decides to share her experiences with her friends and classmates. She uses social media to post about the changes she is making and how they affect her life. She is surprised to receive positive comments and messages from people interested in learning more about how to live more sustainably. Some friends ask her for advice on how to get started, and Alejandra is happy to help them. She realises that sharing her experience not only motivates her to keep going, but can also inspire others to do the same. Thus, without intending to, Alejandra starts to become a voice for sustainability among her acquaintances.

Another initiative Alejandra takes is to promote the use of sustainable transport. Instead of taking her family's car to travel to university, she chooses to use public transport or a bicycle. Although this means getting up a little earlier and, at times, dealing with the Managua heat, Alejandra feels good knowing that she is contributing to reducing carbon emissions. She also enjoys the exercise she gets from cycling and the opportunity to see her city from a different perspective. Each ride becomes an opportunity to reflect on her surroundings and appreciate the small details of everyday life.

Alejandra is also beginning to explore the possibility of producing some of her own food. She uses the small garden at home to grow herbs such as basil, oregano and mint, as well as some vegetables such as tomatoes and peppers. Although space is limited, Alejandra is excited to see how the plants grow and start to bear fruit. The experience of growing her own food connects her even more to the land and makes her appreciate the effort involved in producing food. It also allows her to reduce her dependence on supermarket produce and to have access to fresh, pesticide-free food.

One of the actions that gives Alejandra the most satisfaction is organising a clean-up day in her neighbourhood. She calls some neighbours and friends to pick up the rubbish accumulated in a vacant lot near her house. At first, the response is not as enthusiastic as she had hoped, but with the support of her family and some close friends, they manage to gather a small group. During the day, Alejandra realises that, although the work is hard, the result is worth it. At the end of the day, the land looks clean and the neighbours are grateful for the effort. Alejandra feels great satisfaction in seeing the immediate impact of her actions and knowing that she has contributed to improving her community.

Another important action Alejandra takes is to reduce the consumption of disposable products in her household. She starts replacing paper towels with reusable rags, uses cloth napkins instead of paper and buys products in bulk to avoid excess packaging. These changes take some getting used to, but Alejandra is motivated by the fact that she is reducing the amount of waste she generates. Her family is also starting to adopt these habits, and little by little, everyone is becoming more aware of the impact of their consumption choices.

Alejandra also decides to get involved in environmental education for the children in her community. She organises a small workshop in the local park to teach them about the importance of caring for the environment. She uses games and playful activities to explain concepts such as recycling, saving water and protecting nature. The children participate enthusiastically, and Alejandra is touched to see how young children are willing to learn and care for their environment. She understands that education is key to achieving long-term change and is committed to continuing this type of activity in the future.

In addition, Alejandra begins to reflect on her consumption of clothes and other material goods. She realises that she often buys things she doesn't need and that the fast fashion industry has a big environmental impact. She decides to reduce her clothing purchases and opt for second-hand clothes or clothes from brands that are committed to sustainability. Although at first this represents a change in her lifestyle, she soon discovers that she can dress responsibly without sacrificing her personal style. Alejandra also learns to repair her clothes instead of throwing them away, which gives her a new appreciation for the value of the objects she owns.

6.2 Community and university initiatives

Alejandra faces challenges throughout this process. There are days when she feels frustrated because it seems that her efforts are not enough, especially when she sees other people acting without caring for the environment. However, she remembers the words of Professor Morales: "Every small action counts, and each of us has the power to inspire others". These words motivate her to keep going and not give up, even when things seem difficult. Alejandra understands that change does not happen overnight, but that every step in the right direction is important.

As time goes by, Alejandra begins to notice the positive effects of her actions. Her family has become more conscious of their energy and water consumption, and they have reduced the amount of waste they generate. Her friends have also started to make changes in their lives, inspired by Alejandra's example. Some neighbours have even approached her for advice on how to reduce their environmental impact. Alejandra is proud of what she has achieved and knows that, although there is still a long way to go, she is on the right track.

Alejandra also decides to take her personal actions to the university level. She proposes to her classmates to run a campaign to reduce plastic consumption at the university. Together, they organise activities to raise awareness among students about the impact of single-use plastics and offer sustainable alternatives, such as reusable bottles and cloth bags. The campaign is well received, and Alejandra is encouraged that her efforts are inspiring others to take action.

One of the things Alejandra enjoys most is the constant learning process. Every day she discovers something new about sustainability and finds new ways to apply this knowledge in her life. She realises that living sustainably is not about being perfect, but about doing the best you can with the resources available. Alejandra learns to be more patient with herself and to celebrate every small achievement, no matter how insignificant it may seem. This positive attitude helps her to stay motivated and to keep going, even when she faces obstacles.

Alejandra is also starting to get involved in circular economy initiatives. She is interested in the idea of reusing and recycling materials to give them a second life and reduce the amount of waste that ends up in landfills. Together with some classmates, he decides to organise a creative recycling workshop, where they teach students how to make useful objects from recycled materials. The activity turns out to be a success, and Alejandra is thrilled to see her classmates having fun while learning about the importance of reducing and reusing.

Another initiative Alejandra takes is to reduce the use of chemicals in her home. She starts making her own cleaning products using natural ingredients such as vinegar, baking soda and essential oils. At first, her family is sceptical about the effectiveness of these products, but they soon realise that they work just as well as commercial products, without the negative effects on health and the environment. Alejandra takes satisfaction in knowing that she is contributing to creating a healthier and safer home for her family.

Throughout this process, Alejandra also realises the importance of taking care of herself. She understands that environmental activism can be exhausting and that it is important to find a balance. Alejandra begins to practice yoga and meditation to reduce stress and stay focused on her goals. These moments of reflection help her to reconnect with her purpose and to remember why she started this journey towards sustainability. Alejandra understands that in order to take care of the planet, she must also take care of herself.

6.3 Final reflections and the continued commitment to sustainability

Finally, Alejandra realises that sustainability is not just a set of actions, but a way of life. Every decision she makes, from what she eats to how she travels and what products she consumes, is aligned with her values and her desire to contribute to a better world. Alejandra is proud of the road she has travelled and the changes she has managed to implement in her life. She knows that there is still a long way to go, but she is convinced that every small step counts and that, together, they can make a significant change.

CAPITULO 7

Chapter 7: University Collaboration

Having implemented so many changes in her personal life, Alejandra felt the need to take her actions to a broader level. For her, the university was the ideal place where these projects could have a significant impact. She decided to work together with her peers to develop initiatives that would not only change the university environment, but also inspire other students and academic staff to adopt sustainable practices. This is how he came up with the idea to start collaborating with different student groups and professors interested in sustainability.

7.1 Forming a university sustainability team

Alejandra began by proposing the creation of a formal sustainability team within the university. For this, she met with several of her colleagues, including Lucía and Marcos, with whom she had worked on previous projects. Together, they formed a group dedicated to devising initiatives that promoted sustainability on campus. The team's first meeting was exciting; everyone had ideas and proposals, and the energy in the room was palpable. They discussed projects such as reducing the use of plastic on campus, installing composting stations and creating a community garden where students could learn about sustainable agriculture.

7.2 Initial Projects: Plastic Reduction and Community Garden

One of the first projects they decided to tackle was the reduction of plastic consumption at the university. Alejandra had noticed that many disposable plastic bottles and utensils were being used in the campus cafeterias. With the support of her peers, she met with cafeteria administrators to discuss the possibility of replacing these products with more sustainable alternatives. At first, the administrators were reluctant, as the cost of sustainable products was higher, but Alejandra's team presented a detailed proposal that included an awareness campaign to reduce the use of plastics and encourage the use of reusable bottles.

Prepared by own source

To finance the campaign, the team sought support from the university fund for student projects. Alejandra was in charge of writing the application, explaining the importance of the project and the benefits it would have for the university community. The application was approved, and the team received seed money to start the campaign. With this money, they were able to purchase reusable bottles that were distributed to students during a sustainability fair they organised on campus. The fair included talks, workshops and activities to teach students about the importance of reducing the use of plastics and how their individual actions could contribute to the well-being of the environment.

The sustainability fair was a resounding success. The students actively participated in the workshops and showed great interest in the activities. Alejandra was thrilled to see the response from the university community. Some professors even approached her to congratulate her on the initiative and to offer their support for future projects. This initial success motivated the team to keep going and to think of new projects that could have a positive impact on the campus. It was then that Alejandra came up with the idea of creating a community garden at the university.

The community garden was an ambitious project, but Alejandra was convinced that it could make a big difference on campus. The idea was to have a space where students could learn about sustainable agriculture, grow their own food and connect with nature. To realise this project, the team needed more resources, so Alejandra decided to look for external funding. After researching different options, she found an international fund that supports sustainability projects in educational institutions. Together with her colleagues, she prepared a detailed proposal and submitted it to the fund.

Prepared by own source

Weeks later, Alejandra received an answer that would change the course of the project: the fund had approved their application and would grant them the necessary funding to implement the community garden. The news was received with joy by the whole team, and Alejandra felt proud of what they had achieved. She knew that this was not only a personal achievement, but an important step for the university on its path to sustainability. With the funding secured, they began planning the details of the garden, choosing the right place on campus and defining what types of plants they would grow.

Planning the community garden required collaboration from different areas of the university. Alejandra and her team met with biology professors, who offered their knowledge on best practices for growing and caring for plants. They also worked with architecture students to design the space in an efficient and attractive way. This interdisciplinary collaboration was an enriching experience for Alejandra, as it allowed her to see how people with different skills and knowledge could come together to work towards a common goal. In addition, the garden became a project in which all students could participate and feel part of the process.

As the garden took shape, Alejandra realised that she had created more than just a green space on campus. The community garden was becoming a meeting place, a space where students could disconnect from the stress of their studies and connect with nature and with other students. Alejandra enjoyed every moment she spent in the garden, working in the soil, planting new seeds and watching the plants grow and flourish. With each passing day, the garden became more alive, and with it, the university community became more aware of the importance of sustainability.

The garden project also opened the door to new opportunities. A professor from the environmental science faculty proposed using the garden as an outdoor laboratory where students could conduct research on sustainable agriculture and biodiversity. Alejandra was excited by the idea and started working together with the professor to develop a research programme that could involve students from different majors. The idea was that the garden would not only be a growing space, but also a place where knowledge would be generated and research on environmental issues would be promoted.

In addition to the community garden, Alejandra and her team decided to implement composting stations on campus. They had learned about composting during their visits to communities in Managua and wanted to apply this knowledge at the university. With the support of funding and the collaboration of the campus administration, they set up several composting stations where students and staff could deposit food scraps and other organic waste. The compost generated would be used in the community garden, thus closing the cycle of organic waste and showing the university community how each resource could be put to best use.

The implementation of the composting stations was not an easy task. At first, there was some resistance from some members of the university community, who were unfamiliar with the process and feared that it would generate bad odours or attract pests. To overcome these obstacles, Alejandra and her team organised workshops to teach students and staff how to use the composting stations properly. They also enlisted the help of biology teachers to explain the benefits of composting and how it could contribute to improving soil quality in the community garden.

Prepared by own source

Over time, the composting stations became an integral part of the campus, and more and more people began to use them. Alejandra was pleased to see how her efforts were paying off and how the university community was becoming more aware of the importance of responsible waste management. In addition, the compost generated helped to improve the quality of the soil in the garden, allowing the plants to grow stronger and healthier. Alejandra felt that every step they took brought them closer to their goal of making the university a more sustainable place.

The next project Alejandra and her team decided to tackle was the installation of solar panels on campus. They had learned about the potential of renewable energy during sustainability classes and wanted the university to take a step towards reducing its carbon footprint. To do so, they met with the university administration and presented a detailed proposal that included a cost-benefit analysis and a plan to gradually implement the solar panels in different areas of the campus. Although the project required a considerable investment, Alejandra was convinced that it was worth it.

To their surprise, the university administration showed great interest in the project and decided to support it. In addition, they managed to obtain additional funding from a local organisation promoting the use of renewable energy in educational institutions. With the financial support secured, Alejandra and her team started working on the implementation of the project. They contacted solar energy experts and collaborated with them to install panels on the roofs of some of the campus buildings. Seeing the solar panels installed was a very proud moment for Alejandra and her team, as they knew they were contributing to reducing the environmental impact of the university.

As these projects progressed, Alejandra realised how important it was to have the support of the administration and the different actors within the university. Without their collaboration, many of these projects would not have been possible. Alejandra learned to communicate effectively, to present her ideas clearly and to demonstrate the benefits of each initiative. These skills were fundamental to gaining the necessary support and to successfully carrying out the projects. Furthermore, the experience taught her that sustainability is not just a matter of individual will, but requires the commitment and collaboration of the whole community.

One of the most significant moments for Alejandra was when the university decided to formally recognise the work of the sustainability team. During a ceremony in the main auditorium, the university president congratulated Alejandra and her team for their efforts and presented them with an award for their contribution to the improvement of the campus. Alejandra was excited and proud of what they had achieved. She knew that there was still a long way to go, but this recognition was a sign that they were on the right track and that their efforts were being appreciated by the university community.

The university collaboration also led Alejandra to connect with other students from different universities in Managua. During a conference on sustainability, Alejandra met students from other institutions who were also working on environmental projects. They decided to create a student network for sustainability, where they could share ideas, resources and experiences. This network allowed the students to support each other and enable them to carry out larger and more ambitious projects, not only in their universities, but also in their communities.

Shortly after these achievements, Alejandra received unexpected news: she had been nominated for a scholarship to study abroad, specifically at a university that had an advanced programme in sustainability and renewable energy. The scholarship was offered by an international foundation that supported young leaders committed to environmental change. The nomination came as a surprise, as Alejandra did not know that one of her professors had sent a recommendation highlighting her work and leadership in sustainability projects on campus.

The possibility of studying abroad was something Alejandra had never seriously considered, but the idea filled her with excitement as well as nerves. The scholarship included the opportunity to conduct research in renewable energy and to learn from world-renowned experts. After careful consideration and discussions with her family and friends, Alejandra decided to accept the opportunity. She knew it would be a great challenge, but also a unique opportunity to continue to grow and to gain knowledge that she could then apply back home.

News of the scholarship spread quickly around the university, and Alejandra received support and congratulations from many of her classmates and professors. Although she was excited about what was to come, she was also a little sad to have to temporarily leave the projects she had worked so hard on. However, the sustainability team was willing to continue the work they had started, and Alejandra was confident that they would be able to move forward even in her absence.

Before leaving, Alejandra made sure that all projects were well organised and that each member of the team was clear about their responsibilities. She also pledged to keep in touch with the team and to support where she could from abroad. During a final meeting, she told them how grateful she was for all the support and effort they had put into each initiative. It was an emotional moment for everyone, but also full of hope and a promise to continue working for a more sustainable campus.

Finally, the day came for Alejandra to leave. With a suitcase full of dreams and expectations, she said goodbye to her family and friends, ready to embark on a new adventure. Upon arriving at the university abroad, Alejandra was met with a diverse and stimulating environment. There were students from different countries, all interested in learning about sustainability and finding solutions to global environmental problems. Alejandra felt inspired to be surrounded by people with the same passion for the environment, and knew she was in the right place to continue to grow.

During her studies abroad, Alejandra had the opportunity to learn from renewable energy experts such as Juan Cano and to work on innovative projects that opened her eyes to new possibilities. She participated in research on solar and wind energy, and learned about emerging technologies that could have a big impact on the transition to a more sustainable future. Every day was a new learning experience, and Alejandra felt she was gaining valuable knowledge that she could one day apply in her country.

7.3 Impact and continuity through distance collaboration

Despite the distance, Alejandra kept in touch with the sustainability team at her university in Nicaragua. They organised regular video calls to discuss the progress of projects and to share ideas. Alejandra also shared with them what she was learning in her classes and in her research, and looked for ways to apply that knowledge to the projects she was pursuing on campus. This long-distance collaboration meant that the projects not only remained active, but also continued to evolve.

One of the projects that made the most progress during Alejandra's absence was the installation of more solar panels on campus. Thanks to funding and support from the administration, the team was able to expand the project and cover more buildings with solar panels. Alejandra was proud to see how the campus was becoming a benchmark for sustainability in the region. She knew that her work had been an important part of that process, but she also recognised the effort and dedication of the entire team who had continued to work tirelessly.

After a year of studying abroad, Alejandra returned to Nicaragua full of new ideas and with a renewed determination to continue working for sustainability. Upon arriving at her university, she was welcomed with open arms by her classmates and professors, who were eager to hear all that she had learned. Alejandra organised a series of talks and workshops where she shared her experiences and the knowledge she had gained. The students were enthusiastic and motivated, and Alejandra realised that her journey had not only been an experience of personal growth, but also an opportunity to inspire others.

CAPITULO 8
CENTRO DE INNOV
ENERGIA
SOLAR

Chapter 8: Reflections and Conclusions

8.1 Return to Nicaragua and the Impact of International Experience

fter her experience abroad, Alejandra returned to Nicaragua with an expanded perspective and a wealth of new ideas to implement in her community and at the university. She had learned so much about renewable energy, circular economy and sustainable resource management, and was eager to share this knowledge with others. The trip had not only been an academic experience, but also a personal transformation that had led her to understand the magnitude of environmental challenges and the opportunities that existed to create positive change.

Prepared by own source

One of the first things Alejandra did upon her return was to meet with the university's sustainability team. Everyone was excited to see her and hear about her experiences. Alejandra organised a series of workshops where she shared what she had learned, from new technologies for energy generation to strategies for engaging the community in sustainability projects. The workshops were a success, and many students were inspired by Alejandra's stories about the projects she had worked on and the people she had met during her time abroad.

Alejandra also used her experience to strengthen the student network for sustainability that they had created prior to their departure. She connected this network with some of the contacts she had made abroad, facilitating international collaboration and the exchange of ideas. In this way, the network became a more dynamic and enriching space where students could share their experiences and learn from each other. Alejandra was convinced that collaboration was key to tackling environmental challenges, and seeing how students from different parts of the world were coming together for a common goal filled her with hope.

8.2 Establishment of the renewable energy innovation centre

One of the most ambitious projects Alejandra started to develop after her return was the creation of a renewable energy innovation centre on campus. The idea was to have a space where students could research and develop innovative solutions to the region's energy challenges. Alejandra knew that this would require a lot of work and resources, but she was convinced that, with the support of the university and her peers, they could make it happen. She started by presenting the idea to the university administration, who were interested and willing to support the project.

The innovation centre would be a place where students from different disciplines could work together, sharing their knowledge and skills to develop renewable energy and energy efficiency projects. Alejandra was convinced that practical knowledge was fundamental to finding solutions to environmental problems, and wanted students to have the opportunity to learn by doing. To secure the necessary resources , Alejandra and her team sought external funding, applying to different funds and programmes to support innovation and sustainability.

The planning and fundraising process was long and challenging, but Alejandra never lost motivation. Every small breakthrough, every successful meeting and every contact they made gave them more strength to move forward. Finally, after months of work, they managed to secure the necessary funding to start the project. The innovation centre began to take shape, and Alejandra was excited to see how her vision was becoming a reality. She knew that this would be a space that would benefit not only current students, but also future generations.

8.3 Reflections on education and the role of sustainability in universities

As the innovation centre took shape, Alejandra realised the importance of creating a collaborative and open environment. She didn't want the centre to be just a lab, but a place where students could share ideas, learn from each other and find inspiration for their own projects. To achieve this, they organised a series of events, such as hackathons and innovation competitions, where students could work together to solve problems related to energy and sustainability. These events not only foster creativity, but also help to build a community committed to change.

The innovation centre quickly became a reference point on campus and attracted the attention of students from other universities and external organisations. Professors from different disciplines began to collaborate on projects, and Alejandra was proud to see how the centre was becoming a space where knowledge was generated and research was promoted. More and more students approached her to express their interest in getting involved, and Alejandra was delighted to see how the initial idea had grown into something much bigger than she had imagined.

Prepared by own source

One of the first research projects carried out at the centre was on the use of solar energy in rural communities in Nicaragua. Alejandra had learned a lot about solar technologies during her time abroad and wanted to apply that knowledge to benefit the communities that needed it most. Together with a team of students and teachers, they developed a project to install solar energy systems in a rural community that had no access to electricity. The project was a success, and Alejandra was deeply moved to see how solar energy was improving the quality of life of the people in that community.

8.4 Thinking ahead: Expansion of the model and commitment to sustainability

As the project progressed, Alejandra also began to reflect on the role of education in sustainability. She realised that it was not only important to teach students about sustainable technologies, but also to instil in them a critical mindset and a sense of responsibility towards the environment. To achieve this, she started working with professors from other disciplines to integrate sustainability into the curriculum of different careers. Alejandra was convinced that sustainability should be a cross-cutting theme, present in all areas of knowledge.

Alejandra also used her experience to develop a mentoring programme for students interested in sustainability. She knew that many young people had good ideas, but often did not know how to implement them. The mentoring programme connected students with experts in different areas, who give them guidance and support to develop their projects. Alejandra was happy to see how the programme helped students gain self-confidence and carry out initiatives that had a real impact on their communities.

Over time, Alejandra realised that working on sustainability was an ongoing process. Each project they undertook was a step forward, but also new challenges and opportunities for further improvement. Alejandra learned to see obstacles as opportunities to learn and grow. The experience of studying abroad and leading projects at university had taught her that change did not happen overnight, but that every small effort contributed to a bigger goal.

One of the most significant moments for Alejandra was when the university decided to adopt a long-term sustainability plan, inspired in large part by the projects and work of the sustainability team. This plan included specific goals to reduce the carbon footprint of the campus, promote the use of renewable energy and encourage research on environmental issues. Alejandra was deeply proud to have contributed to this institutional change and to see the university commit to being an example of sustainability for the community.

Alejandra also began working with local organisations and governments to promote public policies that support sustainability. She realised that to achieve real and lasting change, policies needed to support environmental initiatives. Together with other students and civil society organisations, he began to advocate for laws that would encourage the use of renewable energy and the responsible management of natural resources. This experience showed him the importance of citizen participation and working together to influence decisions that affect the environment.

As the projects progressed and the innovation centre consolidated, Alejandra began to think about the future. She knew that there was still much to do, both at the university and in the wider community. Alejandra was committed to continue working for a more sustainable world and to continue inspiring others to do the same. She knew that every little effort counted and that, with everyone's collaboration, they could make a significant change.

The experience of leading the sustainability team and creating the innovation centre had been profoundly transformative for Alejandra. Not only had she learned about sustainability technologies and strategies, but she had also developed leadership, communication and teamwork skills. She had met incredible people, both at home and abroad, and had seen first-hand the power of collaboration in tackling environmental challenges.

Alejandra also began to think about how she could take the innovation centre model to other universities and communities. She was convinced that knowledge should be shared and that the work they had done at her university could serve as inspiration for others. Together with her team, she began to develop a guide on how to create sustainability innovation centres, based on her experience. This guide would be distributed to other universities and institutions interested in promoting sustainability in their communities.

As the innovation centre continued to grow, Alejandra realised that the greatest achievement was not the specific projects they had carried out, but the change in mindset they were bringing about in the university community. More and more students were interested in sustainability and were willing to make changes in their lives to contribute to a better world. Alejandra was deeply gratified to see how her work was inspiring others and how the university was becoming an example of sustainability in the region.

One of the most recent projects Alejandra started to develop was an environmental education programme for local schools. She wanted to take what they had learned at university to the new generations and make sure that children and young people understood the importance of caring for the environment. Together with a team of volunteers, she organised workshops and activities in different schools, teaching students about recycling, energy saving and the importance of biodiversity. Alejandra was thrilled to see the children's enthusiasm and to know that she was helping to shape a new generation of leaders committed to sustainability.

Over time, Alejandra began to reflect on all that she had achieved and the journey she had taken. From her early days at university, when she wondered if she could really make a difference, to now, where she was leading an innovation centre and working with students, teachers and communities to promote sustainability. Alejandra realised that while the challenges were great, so were the opportunities, and that with determination, collaboration and passion, positive change was possible.

Alejandra's journey was not over; in fact, it was just beginning. She was convinced that her work for sustainability would continue, not only at the university, but throughout the country and beyond. She knew that change took time and effort, but she also knew that every step counted. Alejandra was ready to move forward, inspiring others, leading projects and working for a more sustainable future for all.

Conclusion

University Sustainability: Alejandra's Journey showed that sustainability was not just an abstract concept, but a tangible commitment that could be addressed by any individual in their own context. Through Alejandra's experiences, it explored how individual decisions, backed by collective knowledge and effort, could transform environments such as a university or a community into concrete examples of environmentally responsible action.

The story made it clear that personal influence could transcend, becoming a catalyst for systemic change. Alejandra represented the power of committed leadership, demonstrating that small, sustained steps over time could generate significant impact when grounded in ethical principles and a strategic vision.

In these pages, key learnings were synthesised and a thoughtful challenge was posed: to incorporate sustainability as a guiding principle in academic, professional and personal life. The examples and tools presented were not merely theoretical ideas, but practical demonstrations of how every action, no matter how small, can contribute to global goals.

The book concludes with a powerful reminder: change always began with an idea and grew through collaborative action. As these pages close, there remains an unspoken invitation for each reader to reflect on their own role in building a sustainable future and to consider how their choices can make a difference.

In the past, Alejandra began this journey with questions and uncertainties; in the present, her story left a transformative lesson: every action has the potential to shape the future. What place will the reader have in this collective narrative? The future, as this book suggests, will always be determined by the decisions of those who dare to act.

References

Borrello, M., Caracciolo, F., Lombardi, A., Pascucci, S., & Cembalo, L. (2017). Consumers' perspective on circular economy strategy for reducing food waste. *Sustainability, 9*(1), 141. https://doi.org/10.3390/su9010141. https://doi.org/10.3390/su9010141

Ceballos, G., Ehrlich, P. R., & Dirzo, R. (2017). Biological annihilation via the ongoing sixth mass extinction signaled by vertebrate population losses and declines. *Proceedings of the National Academy of Sciences, 114*(30), E6089-E6096. https://doi.org/10.1073/pnas.1704949114

Energy Information Administration (EIA). (2020). Renewable energy in Central America. https://www.eia.gov/renewable/

García, A. (2019). The importance of environmental education in the university context. *Journal of Sustainable Education, 12(*3), 112-125.

Geistdörfer, M., Savaget, P., Bocken, N. M. P., & Hultink, E. J. (2017). The circular economy - A new sustainability paradigm? *Journal of Cleaner Production, 143*, 757-768. https://doi.org/10.1016/j.jclepro.2016.12.048

Goldstein, N. (2021). Composting and its impact on urban communities. *Environmental Science & Policy, 120*, 98-104.

Intergovernmental Panel on Climate Change (IPCC). (2018). Global warming of 1.5°C. https://www.ipcc.ch/sr15/

International Renewable Energy Agency (IRENA). (2020). Renewable energy statistics 2020. https://www.irena.org

Lovins, A. B. (2018). Energy efficiency: The essential foundation of a sustainable energy future. *Energy Efficiency Journal, 11*(4), 993-1005.

Morales, R. (2019). Application of the circular economy at the university level. *Journal of Environmental Studies, 8*(2), 45-62.

United Nations (2015). Sustainable Development Goals. https://www.un.org/sustainabledevelopment/es/

World Health Organization (WHO). (2018). Climate change and health. https://www.who.int/es/news-room/fact-sheets/detail/climate-change-and-health

Senge, P. (2006). *The fifth discipline: The art and practice of the learning organisation*. Granica.

United Nations Environment Programme (UNEP) (2020). Youth and the environment: Engaging young people in sustainability. https://www.unep.org/youth-sustainability

Andrews, L. (2018). *Educating for sustainability: Strategies to inspire change in university curricula*. Springer.

Bocken, N. M. P., Short, S. W., Rana, P., & Evans, S. (2014). A literature and practice review to develop sustainable business model archetypes. *Journal of Cleaner Production, 65*, 42-56. https://doi.org/10.1016/j.jclepro.2013.11.039

Brown, L. R. (2019). *The great transition: Shifting from fossil fuels to solar and wind energy*. Earthscan.

Costanza, R., Fisher, B., Ali, S., Beer, C., & Bond, L. (2007). Quality of life: An approach integrating opportunities, human needs, and subjective well-being. *Ecological Economics, 61*(2-3), 267-276. https://doi.org/10.1016/j.ecolecon.2006.02.023

Densham, M., & Jones, L. (2017). *Understanding renewable energy: From technology to policy*. Routledge.

Elkington, J. (1999). *Cannibals with forks: The triple bottom line of 21st century business*. Capstone Publishing.

García, L., & Sánchez, A. (2020). The transition to renewable energy in Latin America: Challenges and opportunities. *Journal of Energy and Climate Change, 15*(4), 89-103.

George, A. (2015). *Energy independence: Achieving sustainable energy for the future*. University Press.

Global Environmental Facility (GEF). (2021). Climate change and renewable energy solutions. https://www.thegef.org/topics/climate-change

Goleman, D. (2009). *Emotional Ecology: The impact of our emotions on the environment*. Editorial Kairós.

Jørgensen, M. S., & Remmen, A. (2015). Circular economy: Conceptual clarifications and its application to the energy transition. *Journal of Sustainable Development, 12*(1), 115-127. https://doi.org/10.1080/21683565.2015.1051437

Kapoor, A., & Kaul, R. (2019). Circular economy in the context of renewable energy and environmental sustainability. *Renewable and Sustainable Energy Reviews, 107*, 324-331. https://doi.org/10.1016/j.rser.2019.02.037

Krzywoszynska, A. (2018). Urban sustainability and the circular economy. *Environmental Sociology, 4*(2), 204-220. https://doi.org/10.1080/23251042.2018.1474190

Lacy, P., & Rutqvist, J. (2015). *Waste to wealth: The circular economy advantage*. Palgrave Macmillan.

Larsen, M., & Brunstad, O. (2019). *Sustainability and the global economy: The role of renewable energy*. Academic Press.

Lele, S. (2018). *Building sustainable institutions for global renewable energy development*. Wiley-Blackwell.

OECD (2020). Circular economy and the environment: An overview. *OECD Environment Policy Papers, 20. https://www.*oecd.org/environment/circular-economy/

Perez, C., & Oliveira, R. (2019). The role of sustainability education in university curricula: Developing an interdisciplinary approach. *Journal of Environmental Education, 50*(1), 60-74. https://doi.org/10.1080/00958964.2018.1521442

Sachs, J. (2015). *The age of sustainable development*. Columbia University Press.

Schmidt, K., & Vachon, M. (2017). Business strategies for sustainability in renewable energy industries. *Environmental Impact Assessment Review, 64*, 49-55. https://doi.org/10.1016/j.eiar.2017.03.004

United Nations Framework Convention on Climate Change (UNFCCC) (2021). Renewable energy and climate action. https://unfccc.int/topics/renewable-energy

Wilson, M., & James, D. (2016). *Environmental policy: New directions for sustainable development*. Cambridge University Press.

MIX
Papier aus verantwortungsvollen Quellen
Paper from responsible sources
FSC® C105338

Printed by Books on Demand GmbH, Norderstedt / Germany